ALTERNATOR BOOKS™

TECH·TITANS

THE GENIUS OF MICROSOFT

How Bill Gates and Windows Changed the World

I0786901

Margaret J. Goldstein

Lerner Publications ◆ Minneapolis

Names and trademarks appearing in this book are the property of Microsoft Corporation, and are used solely for identification purposes. Alternator Books is a registered trademark of the Lerner Publishing Group. This book is a publication of the Lerner Publishing Group, and it has not been approved, endorsed or licensed by Microsoft Corporation.

Copyright © 2023 by Lerner Publishing Group, Inc.

All rights reserved. International copyright secured. No part of this book may be reproduced, stored in a retrieval system, or transmitted in any form or by any means—electronic, mechanical, photocopying, recording, or otherwise—without the prior written permission of Lerner Publishing Group, Inc., except for the inclusion of brief quotations in an acknowledged review.

Lerner Publications Company
An imprint of Lerner Publishing Group, Inc.
241 First Avenue North
Minneapolis, MN 55401 USA

For reading levels and more information, look up this title at www.lernerbooks.com.

Main body text set in Aptifer Sans LT Pro.
Typeface provided by Linotype AG.

Library of Congress Cataloging-in-Publication Data

Names: Goldstein, Margaret J., author.
Title: The genius of Microsoft : how Bill Gates and Windows changed the world / Margaret J. Goldstein.
Description: Minneapolis, MN : Lerner Publications Company, an imprint of Lerner Publishing Group, Inc., [2022] | Series: Alternator books. Tech titans | Includes bibliographical references and index. | Audience: Ages 8–12. | Audience: Grades 4–6. | Summary: "With its groundbreaking Windows operating system, Microsoft transformed personal computing. Learn how founder Bill Gates turned Microsoft into one of the biggest companies in the world using cool technology and the power of coding"— Provided by publisher.
Identifiers: LCCN 2021021728 (print) | LCCN 2021021729 (ebook) | ISBN 9781728440835 (lib. bdg.) | ISBN 9781728449548 (pbk.) | ISBN 9781728445274 (eb pdf)
Subjects: LCSH: Windows (Computer programs)—Juvenile literature. | Microsoft Corporation—Juvenile literature. | Gates, Bill, 1955—-Juvenile literature.
Classification: LCC QA76.76.W56 G6485 2022 (print) | LCC QA76.76.W56 (ebook) | DDC 005.4/37—dc23

LC record available at https://lccn.loc.gov/2021021728
LC ebook record available at https://lccn.loc.gov/2021021729

Manufactured in the United States of America
1 – CG – 7/15/22

TABLE OF CONTENTS

INTRODUCTION

Bill Gates' computer skills helped make him one of the wealthiest people on Earth.

Bill Gates was confident. The teen from Seattle, Washington, was still in high school, but his skills were in demand. He had mastered several computer languages, including BASIC and COBOL. He knew how to program computers. At that time, in the early 1970s, very few people had that ability.

Bill had already put his skills to the test several times. Along with friends in the Lakeside School computer club, he had tested a mainframe computer for a team of engineers from the University of Washington. The engineers wanted the Lakeside club to find bugs in their code and gave the students free computing time in exchange for their work.

After successfully completing that project, Lakeside hired Bill to computerize the school's class scheduling system. Next, Bill and fellow computer club member Paul Allen formed a company called Traf-O-Data. They designed a computer system for analyzing traffic patterns for city governments.

Bill believed that his computer skills would make him rich. He imagined running a big computer company. But the high schooler had no idea just how big his business would become.

A young Gates (*right*) and Paul Allen at Lakeside School in 1970

Harvard Square in the 1970s

CHAPTER 1:
THE BEGINNING

After high school, Gates enrolled at Harvard University in Cambridge, Massachusetts. He majored in math. Paul Allen, who was two years older than Gates, studied at Washington State University for a couple years before taking a programming job with the Honeywell Corporation. The job was in Boston, near Cambridge, so the two friends remained close.

Because of advances in microprocessors, computers were getting smaller. Gates and Allen believed that someday, everyone would have a personal computer at home. They talked about forming a company to write software for personal computers.

The Altair 8800, one of the first personal computers

THE ALTAIR BREAKTHROUGH

In early 1975, *Popular Electronics* magazine ran a story about a kit computer, a small machine that buyers assembled themselves. Called the Altair 8800, it came from a company called MITS in Albuquerque, New Mexico. Computer enthusiasts were excited about the Altair. It signaled the beginning of the personal computer industry.

When they learned about the Altair, Gates and Allen saw a business opportunity. The computer needed software, and the two friends set out to write it. They designed a program that would allow users to run BASIC on the Altair.

When the code was finished, Allen flew to Albuquerque to show it to Ed Roberts, the head of MITS. Roberts agreed to sell the software with his machines. Allen and Gates licensed their software to MITS using the company name Micro-Soft. Allen also moved to Albuquerque and took a job with MITS.

Several times, Gates took time off from Harvard to join Allen in New Mexico. There, he wrote more code for the Altair. In November 1976, Allen left his MITS job. He and Gates officially registered their software business with the state of New Mexico. This time, they spelled the name *Microsoft*.

WHAT IS CODE?

A code is a set of instructions for a computer. To write code, programmers use computer languages, such as Java or Python. Computer languages consist of words, letters, numbers, and symbols with different meanings. They tell computers which electronic circuits to switch on or off to perform calculations and tasks.

THE PC EXPLOSION

In the winter of 1977, Gates dropped out of Harvard. He moved to Albuquerque to run Microsoft with Allen. The company created software for MITS and other technology businesses. In 1979, Gates and Allen moved the company to the Seattle area, where they had grown up.

Personal computers were at last becoming a reality. With small machines from Apple and other companies, users could create and edit documents, keep track of

The IBM PC

finances, and play games. In 1981, the business machine company IBM built the IBM Model 5105, commonly called the IBM PC. IBM hired Microsoft to provide an operating system (OS) for this computer. It was called MS-DOS.

IBM was a wealthy and well-known company. Many buyers chose the IBM brand when they purchased their first home computer. The IBM PC quickly outsold machines from Apple and other computer makers. Some companies built "IBM clones," which were nearly identical to IBM's machine. All the clones ran MS-DOS, just like the IBM PC.

MS-DOS became the most widely used OS, and Microsoft's sales skyrocketed. The company hired thousands of staffers, opened new offices, and created new kinds of software. Allen left the company in 1983 to pursue other projects. Gates remained as chief executive officer (CEO).

MS-DOS was short for Microsoft Disk Operating System. The OS was stored on floppy disks.

INDUSTRY STANDARD

Microsoft kept improving its software. To work with MS-DOS, users had to type in codes and commands. Microsoft's next OS, Windows, was more user-friendly. Introduced in the mid-1980s, it featured mouse-clickable icons and easy-to-use pull-down menus. The company also created Word for word processing, Excel for making spreadsheets, and other popular software.

Windows 95 came out in 1995. It introduced the popular Start menu that came to define Windows.

A Windows start screen on a Microsoft Surface tablet

In the 1990s, at the start of the internet age, Microsoft created the Internet Explorer web browser. The company continued to dominate the personal computing industry. By the year 2000, about 97 percent of all personal computers sold around the world ran the Windows OS, along with other Microsoft programs.

Customers check out Surface laptops and other Microsoft products at a Microsoft store in New York City.

CHAPTER 2:
STANDOUT PRODUCTS

Microsoft remains the world leader in computer software. Windows is still the most widely used OS for personal computing. Around the globe, more than one billion people use Microsoft Office, a group of apps including Word, Excel, and PowerPoint. In addition, Microsoft offers the Microsoft Edge web browser and the Bing search engine. Microsoft also sells hardware, including keyboards, mice, gaming devices, and Surface laptop and tablet computers.

SPECIAL EFFECTS

In addition to Office, Microsoft offers programs for mapping, database management, cybersecurity, language translation, web conferencing, and many other tasks. It builds software for specific industries, such as health care, retail, and manufacturing. It makes additional tools for coders, website developers, and other information technology professionals.

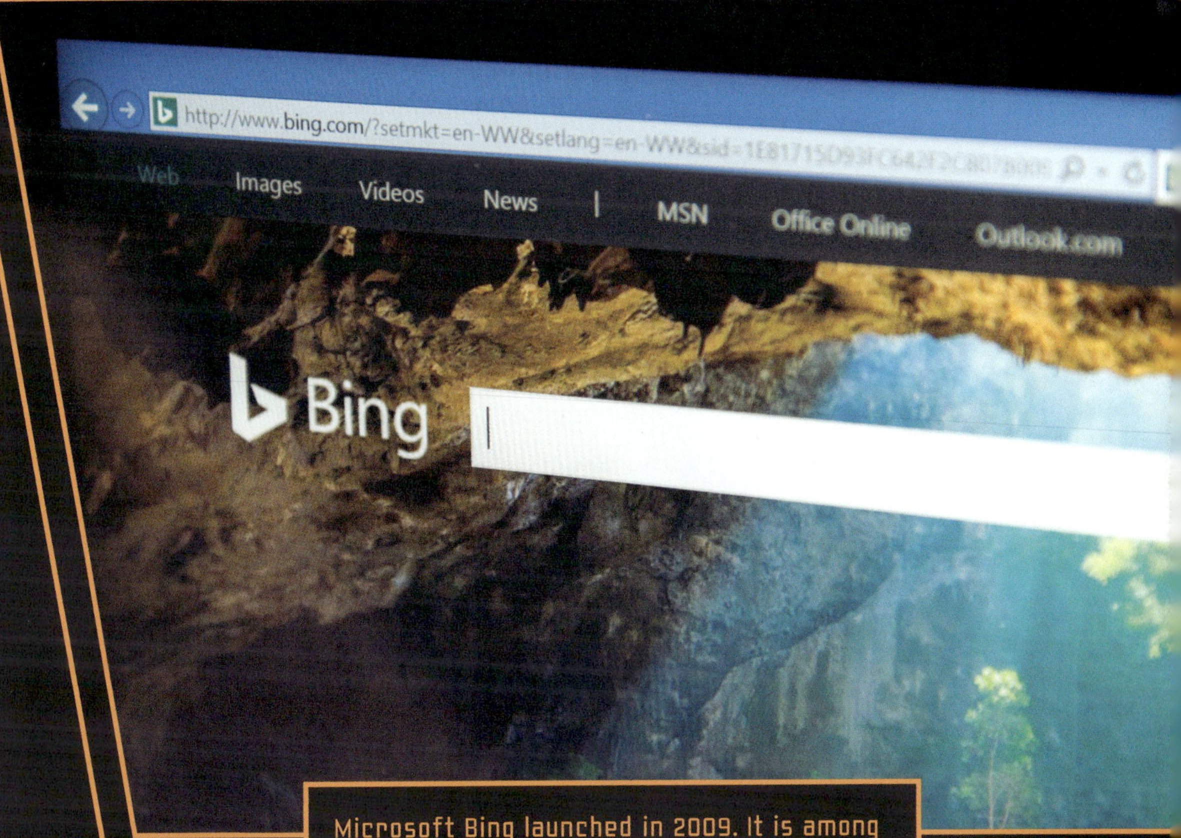

Microsoft Bing launched in 2009. It is among the world's top search engines after Google.

Xbox is Microsoft's gaming division. It sells games, gaming consoles, and accessories. One of the most popular Xbox games is Minecraft, where players can build, explore, and create in an imaginary world.

Microsoft owns several businesses that operate under their own brand names. One of them is Skype, which lets users make video and voice calls to one another. Microsoft also owns LinkedIn, a social media site for business networking, job searches, and career building.

The Xbox Series X (*black*) and Xbox Series S (*white*) came out in November 2020. The Series X comes with a disk drive for games and movies, while the Series S is an all-digital console.

BIG NAMES IN GAMING

The most famous names in gaming are Nintendo, PlayStation from Sony, and Xbox from Microsoft. Nintendo, a Japanese company, began in the late 1800s. It first made playing cards. In the twentieth century, Nintendo started making electronic toys, arcade games, and home gaming consoles. It introduced its popular *Super Mario Brothers* game in 1985. The electronics company Sony also began in Japan. It launched its first video console, the PlayStation, in 1994.

Microsoft got into the gaming world with the Xbox in 2001. It was the first company to offer online gaming. In 2021, PlayStation dominated the gaming market. More than half of all gaming consoles sold were PlayStations. Xbox came in second, with more than 40 percent of console sales. Nintendo had faded, with less than 1 percent of the gaming console market in 2021.

PlayStation 5 console and controller

The first Mac OS came out in 1984. Above, an iBook computer runs on Mac OS 9, which came out in 1999.

CHAPTER 3:
IN THE NEWS

From the start, other computer businesses criticized Microsoft. They said it didn't play fair. In 1988, Apple sued Microsoft for copyright infringement. Apple said Microsoft had copied its user-friendly Macintosh OS to create Windows, without getting permission or paying Apple for the design. A judge decided in favor of Microsoft, ruling that most of the design ideas Microsoft had copied weren't protected by copyright laws.

To beat the competition, Microsoft bought smaller software businesses or incorporated their best ideas into its own software. It also forbade some computer makers to sell competitors' software with Windows-based machines. To make Internet Explorer the world's leading web browser, Microsoft created code that made Explorer work perfectly with Windows. But using another web browser with Windows was difficult. The US Justice Department said this practice was illegal and in 1998 filed charges against Microsoft. The case ended with Microsoft agreeing to stop this and other anticompetitive practices.

Gates defended Microsoft's practices before Congress in March 1998.

THE GREAT GATES

Teenage Bill Gates had dreamed of being rich—and that dream came true. He was a billionaire by age thirty-one. In 1995, with his personal wealth totaling $12.9 billion, Gates became the world's richest man. He has held that title on and off since then.

Gates wanted to use his money to make positive changes in society. In 1994, he and his wife launched a charitable foundation. The Bill & Melinda Gates Foundation carries out health, education, and social justice projects around the world, especially in poor nations.

In 2000, Gates stepped down as Microsoft CEO. However, he remained on the company's board of directors, first as board chair and then as a technology adviser. In 2020, he resigned from the board to devote more time to the Gates Foundation.

Bill and Melinda Gates speak at the 2018 Goalkeepers event honoring leaders who have worked to address major global issues, such as poverty and inequality.

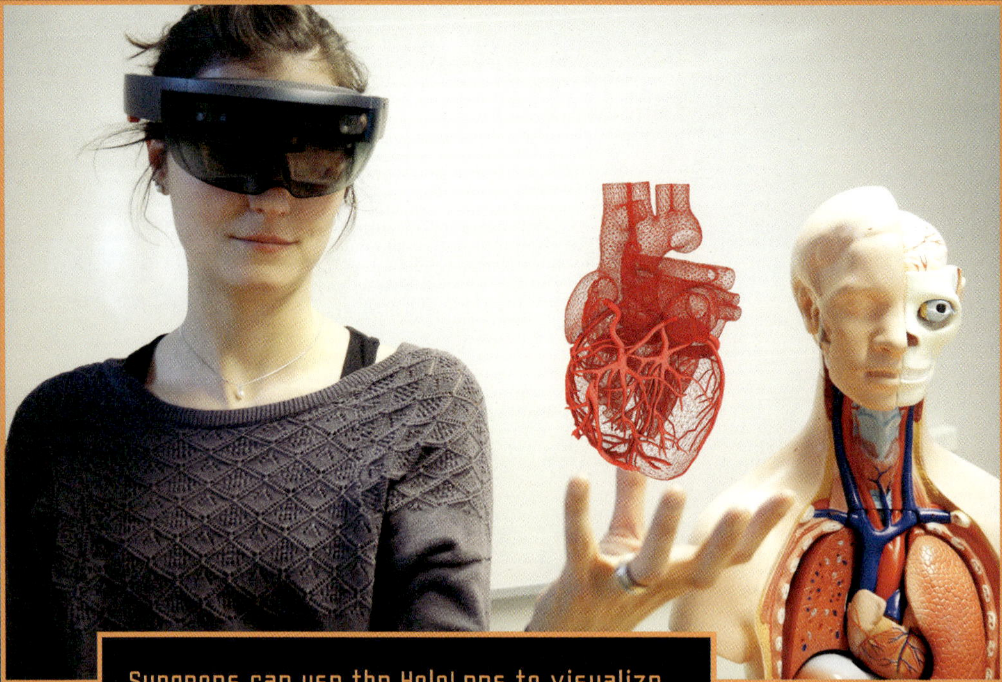

Surgeons can use the HoloLens to visualize organs and prepare for operations.

CHAPTER 4:
A LOOK AHEAD

Microsoft remains at the cutting edge of computer technology. Its most exciting products immerse users in virtual and mixed reality. With the HoloLens headset, for example, users can see holograms layered over their real-world vision. They can interact with virtual menus, maps, diagrams, and 3D images using hand gestures and voice commands. Organizations use the HoloLens to train soldiers, doctors, engineers, and other professionals. Gamers use it to immerse themselves in 3D worlds.

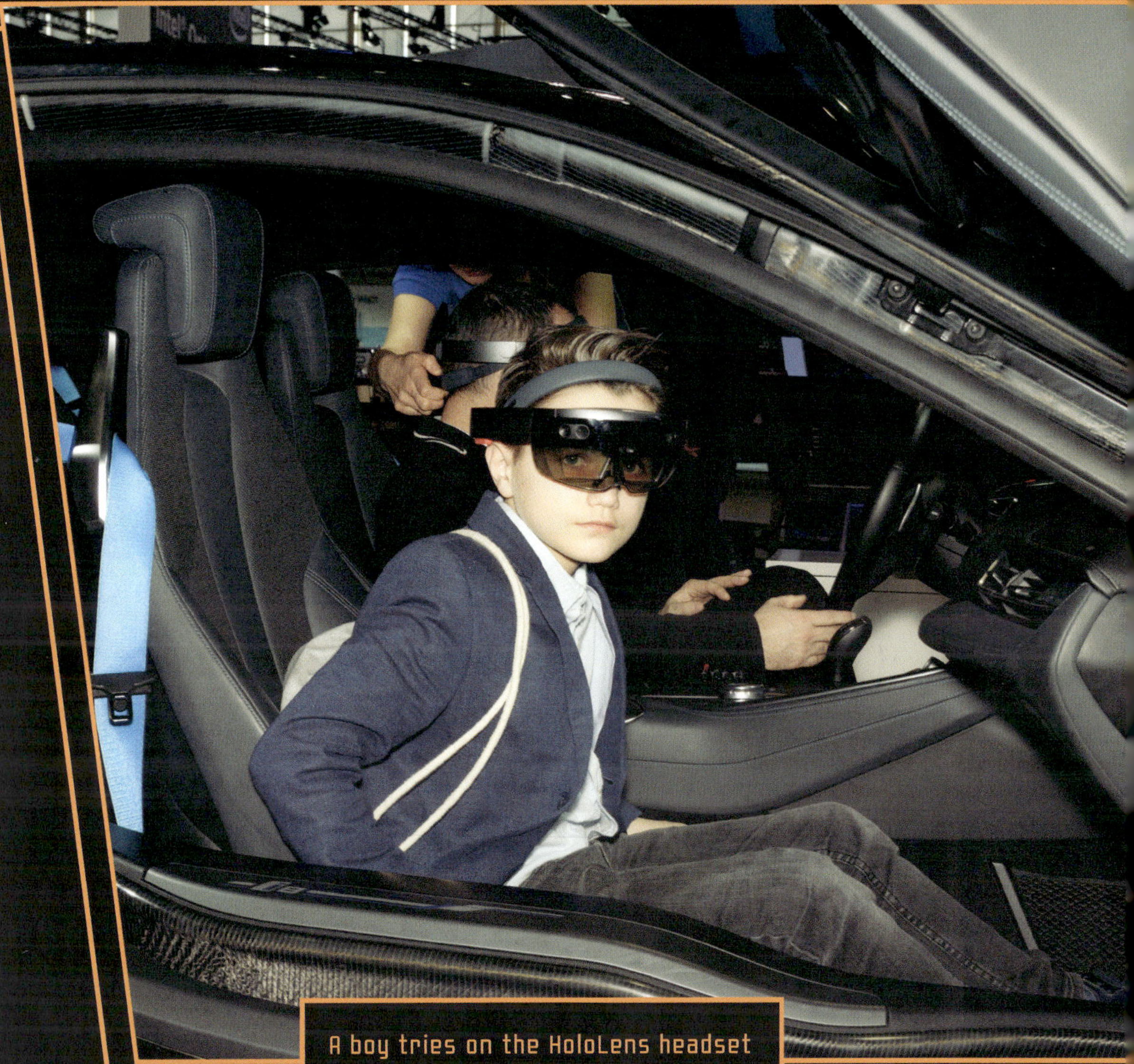

A boy tries on the HoloLens headset at an information technology trade fair in Germany.

WHAT IS HOLOGRAPHY?

Creating holograms involves lasers, which are extremely powerful beams of light. A laser beam travels through a lens, bounces off a mirror, and shines on the object to be pictured. The light bounces again, transferring the image of the object to light-sensitive material, such as a photographic plate. Another laser beam shines on the plate, and the two beams intersect. The meshing of the two beams creates an image that looks 3D.

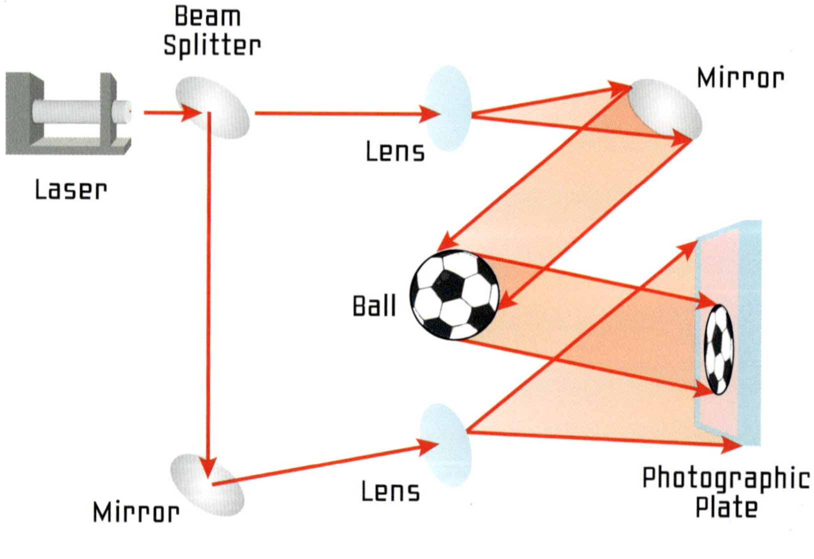

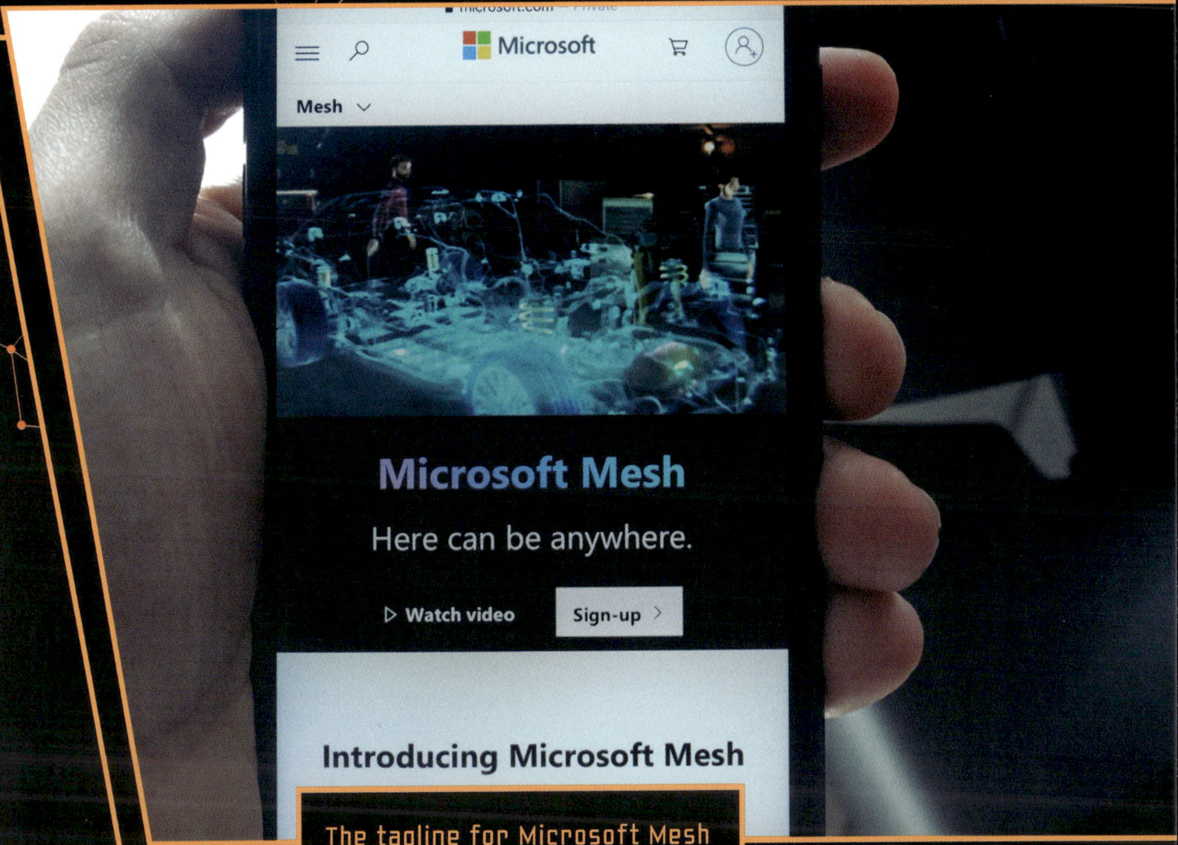

The tagline for Microsoft Mesh is "Here can be anywhere."

Microsoft Mesh takes the HoloLens system to a new level. It allows users in different places to interact in a holographic world. People using Microsoft Mesh don't just connect with virtual text and images. They actually appear to one another as holograms, whether they are standing in the next room or halfway around the world. Everyone shares a virtual space and can work on the same hands-on virtual project.

POWER SHIFT

Like Gates, Microsoft's new leaders, including CEO Satya Nadella, are committed to fighting climate change. Microsoft has set a goal to be carbon negative by 2030, meaning it hopes to emit less carbon than it uses. Microsoft has already started making its company headquarters in Redmond, Washington, carbon negative. It is building a system to generate electricity using solar power and heat from deep inside Earth.

Satya Nadella first joined Microsoft as an engineer in 1992. He held numerous leadership roles before being named CEO in 2014.

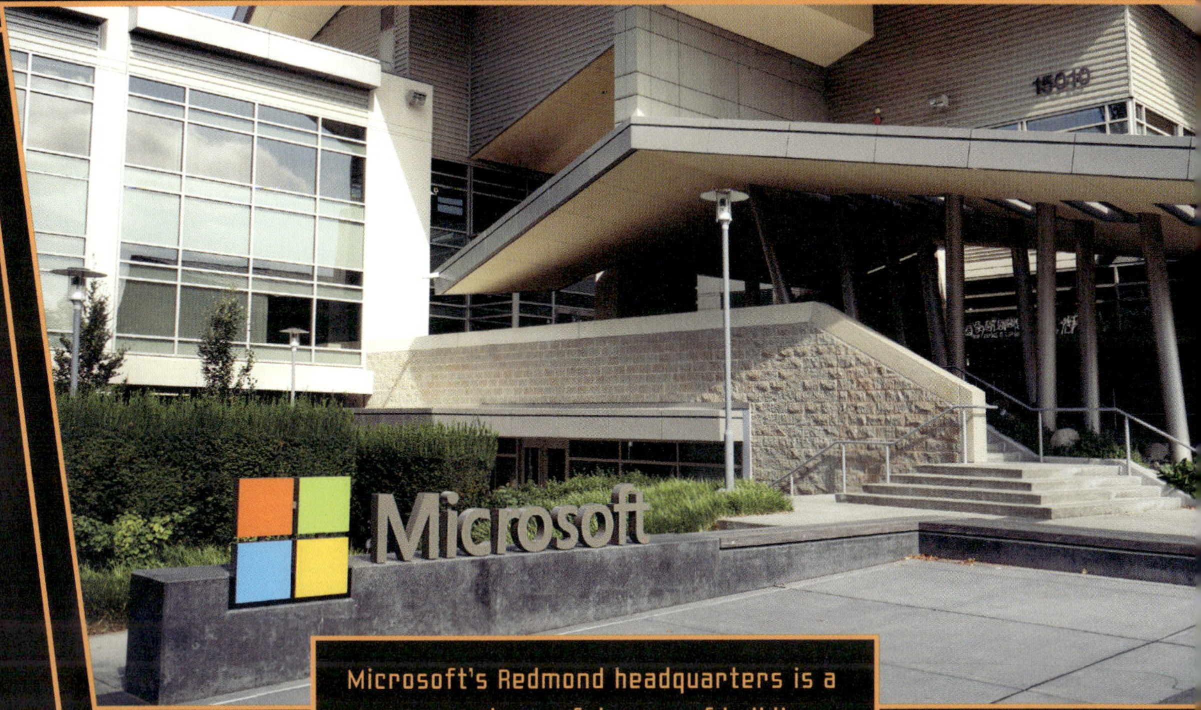

Microsoft's Redmond headquarters is a campus made up of dozens of buildings. It even has its own transit system!

The company is also investing in reforestation, or tree-planting projects, because trees naturally absorb carbon from the atmosphere. And, it has teamed up with automaker General Motors to develop electric cars. Unlike gasoline-powered cars, electric cars do not emit gases that contribute to climate change. Microsoft is committed to using innovation of all kinds to build a better future.

TIMELINE

1976: Bill Gates and Paul Allen register their software business, named Microsoft, with the state of New Mexico.

1979: Microsoft moves from Albuquerque to the Seattle area.

1981: IBM introduces its first personal computer, which is equipped with Microsoft's operating system, MS-DOS.

1985: Microsoft releases its first version of the Windows operating system.

1994: Gates and his wife, Melinda, create the Bill & Melinda Gates Foundation.

1995: Microsoft releases the web browser Internet Explorer.

1998: The US Justice Department rules that Microsoft uses illegal tactics to defeat competing companies.

2000: Bill Gates resigns as Microsoft CEO.

2001: Microsoft launches the Xbox gaming console.

2014: Satya Nadella becomes Microsoft's CEO.

2020: Microsoft announces plans to be carbon negative by 2030.

GLOSSARY

bug: a flaw in a computer code or program

carbon negative: taking more carbon out of the atmosphere than the amount released into the atmosphere

climate change: rising temperatures on Earth created by the burning of fossil fuels, which releases heat-trapping carbon dioxide into the atmosphere

copyright: laws that prevent people from copying, distributing, or profiting from the creative work of others without permission

hologram: an image on a flat surface that appears to be 3D

license: to allow another person or organization to use software or another product, usually for a fee

mainframe computer: a large computer designed to handle large amounts of data

microprocessor: the electronic circuitry for running a computer, contained on a tiny piece of material, such as a silicon chip

operating system (OS): software that controls programs, memory, and the flow of information into and out of a computer

personal computer: a small computer designed to be operated by one person

LEARN MORE

Bill Gates, Pioneer of Modern Computers
https://easyscienceforkids.com/all-about-bill-gates/

Make Your Own Virtual Reality Glasses
https://www.instructables.com/Make-Your-Own-Virtual-Reality-Glasses/

Martin, Chris. *Minecraft: The Business Behind the Makers of Minecraft*. Minneapolis: Lerner Publications, 2016.

Microsoft Coding for Kids
https://techcommunity.microsoft.com/t5/small-basic-blog/microsoft-coding-for-kids/ba-p/337985

Oachs, Emily Rose. *The Personal Computer*. Minneapolis: Bellwether Media, 2019.

Paul Allen Facts for Kids
https://kids.kiddle.co/Paul_Allen

Preuitt, Sheela. *Mission Python*. Minneapolis: Lerner Publications, 2020.

Roza, Greg. *Bill Gates: Founder of Microsoft*. New York: PowerKids Press, 2016.

INDEX

PHOTO ACKNOWLEDGMENTS

The images in this book are used with the permission of: © Frederic Legrand - COMEO/Shutterstock Images, p. 4; © Bruce Burgess/Wikimedia Commons, p. 5; © Fred Jellison Jr./Wikimedia Commons, p. 6; © Michael Holley/Wikimedia Commons, p. 7; © SeventyFour/Shutterstock Images, p. 9 (IT developer); © Casimiro PT/Shutterstock Images, p. 9 (code); © Jordiferrer/Wikimedia Commons, p. 10; © Project kei/Wikimedia Commons, p. 11; © Roland Magnusson/Shutterstock Images, p. 12; © Kārlis Dambrāns/Wikimedia Commons, p. 13; © Anton Gvozdikov/Shutterstock Images, p. 14; © Pe3k/Shutterstock Images, p. 15; © ALDECA studio/Shutterstock Images, p. 16; © Mohsen Vaziri/Shutterstock Images, p. 17; © Felix Winkelnkemper/Wikimedia Commons, p. 18; © Rob Crandall/Shutterstock Images, p. 19; © Greg Allen/AP Images, p. 21; © Valery Heritier/Flickr, p. 22; © flowgraph/Shutterstock Images, p. 23; © Mighty Media, Inc., p. 24; © Justlight/Shutterstock Images, p. 25; © Moritz Hager/Flickr, p. 26; © vladdon/Shutterstock Images, p. 27.

Cover Photos: © Frederic Legrand - COMEO/Shutterstock Images (Bill Gates); © hafakot/Shutterstock Images (Microsoft research and development center); © McLittle Stock/Shutterstock Images (smartphone)

Design Elements: © Hluboki Dzianis/Shutterstock Images